I0467292

Introduction

Herbs have been an important part of human lives. They are used in a lot of dishes to add the aroma to the meals. We often come across recipes and dishes that ask for fresh Parsley, Basil or Chives and we end up using the dried herbs available in the market which lacks the freshness and the aroma expected of it. Now, wouldn't it be better if you could have the same herbs freshly at your home? At a cost much lower than what you pay in market? Well, we've got a solution for this. You can now grow herbs at home at much less efforts and cost. You can grow almost any kind of herb at home like Chives, parsley, Basil, Cilantro, Thyme and much more. All these plants although require slightly different conditions to grow, however they are not difficult to provide because it does not need much space to grow. There are a lot of benefits of growing your own herbs at home, but one of the major positive points is that the herbs are completely healthy and free of all kinds of additives, chemical sprays and other genetically modified ingredients that may affect your health.

To grow herbs at home you will need to carefully set a specific portion for them at home with enough air supply. Herbs need little but regular water supply to grow in. Similarly, you will have to make sure that the herbs get enough light to get their photosynthesis properly. This way you will be able to grow natural herbs at home which would be inexpensive, convenient and healthy choice for your regular consumption. The book in your hands will pick you from scratch and teach you all that you need to know in order to successfully grow herbs at your home. So read it and start growing natural herbs at your home conveniently.

Contents

Chapter 1

The Benefits Of Growing Your Own Herbs

Chapter 2

The Basics Of Growing Herbs In Containers

Chapter 3

Useful Tools

Chapter 4

The Potting Soil

Chapter 1

The Benefits Of Growing Your Own Herbs

With the availability of a wide variety of fresh and dried herbs these days at your local supermarket or grocer, why would you bother growing your own herbs? Why go to the trouble of getting your hands dirty and waiting patiently for your produce to grow when you can just open a packet and get the same basic ingredients? Well, here are only a few of the many benefits you can enjoy from growing, rather than buying, your own herbs. I'm sure after you start growing your own herbs you'll find your own reasons.

1.You will save money- Fresh herbs, especially, can be very expensive when you buy them from your local supermarket and most people, once they have used what they need, will let the leftover herbs rot. Most herbs, once established, require little work or maintenance.

2.Home grown herbs are a healthier option.- As well as the known medicinal benefits of herbs, did you know that adding herbs to your food is a great way to boost your vitamin intake?

3.It's an opportunity to learn- There's so much to learn about herbs, when you have your own ready supply of herbs, you will find new and wonderful ways to use your abundant supply of herbs.

4.You always have a convenient supply of fresh herbs- No more running to the shops to buy your herbs to only find they don't have what you're looking for or trying to find a shop that has the herbs you need. One of the greatest advantages of growing your own herbs is always having the herb you need right at your fingertips- ready to add flavour to any food you're cooking.

5.Herbs add variety- Your local shops will usually stock no more than one type of any particular herb. Did you know that there are many different types of the same basic herb? For example, there are over thirty types of basil. Growing your own herbs will allow you to sample many exotic new herbs and discover a whole new world of flavours.

6.Herbs can enhance the look of your house- Most herbs are a just as attractive as flowers and shrubs. Many herbs have a pleasing aroma which can help beautify your home and your home's ambiance.

7.It can reduce stress- Tending to or visiting your herbs has a calming effect and helps to relieve all that built up stress that modern living gives us. The sights and smells of your own potted herbs revitalises the soul and delights the senses.

8.You can share or sell any extra herbs you have- When you grow your own herbs you will always have more herbs than you will know what to do with, which will leave you with plenty left over. You can give these leftovers away to friends or family or sell for a profit.

9.Herbs can transform even your most basic meals- Adding a few herbs to a simple pasta or meat dish will completely transform the meal. Basic meals can be a new adventure on a nightly basis. The possibilities are only limited by your imagination

10.Herb Gardening is enjoyable hobby- The satisfaction of seeing what you planted grow is a wonderful feeling. It helps you take your mind of your troubles. With a few skills, herbs are so easy to grow and maintain; there is little chance for disappointment.

11.Potted herbs make great gifts- If placed in an attractive pot, herbs can make useful and attractive gifts.

Chapter 2

The Basics Of Growing Herbs In Containers

Almost any herb can easily be grown in a container. However some herbs may have different water requirements and some herbs are more particular in their watering requirements than others so it's a good idea to place herbs that have the same maintenance requirements together in the same pots.

Choosing what herbs to plant- Plant any herbs you like the look, taste and smell of. Herbs such as rosemary make good candidates because depending on where you live, rosemary can be expensive to buy can be used in a wide variety of dishes and once established, requires very little maintenance.

Soil Conditions- Most herbs require good drainage so only use high quality potting soil and always insure that you plant your herbs in a container that has adequate drain holes so you don't drown your herbs.

Fertiliser- Make sure you don't over-feed your herbs. Most herbs don't require much fertiliser and over-fertilising many types of herbs may kill them. Some herbs, such as oregano and thyme, don't require much work once established and will not taste as good if they are given too much food, water or attention.

Choosing the right container for your herbs- Most herbs can be planted in any container as most herbs don't have larger root systems and some tolerate drying out between waterings. However keep in mind that the less soil your pot contains per plant, the less margin of error you have when it comes to watering your plants.

Light- Most herbs will require at least six hours of full sun a day. However containers can bake during a very hot day, especially if you live in a hotter climate. It may be a good idea to store your potted herbs in a shaded area during the hotter part of the day. Most herbs can be placed indoor under a sunny window sill.

Multi-plant containers- Many smaller herbs can do well when planted next to other herbs in the same large container. If you periodically prune them, you can prevent them from competing with each other for light. However one plant that doesn't like being in the same pot as other herbs is basil as basil requires good air circulation.

Self-Watering pots- Some herbs such as parsley, mint and chives can do well in self-watering containers because they like to have a constant level of moisture. However, other herbs such as rosemary, oregano and basil prefer dryer conditions and therefore herbs such as these should not be planted in self-watering containers .

Deciding which herbs to combine in the one container- Generally you can grow as many types of herbs as you like in one container provided they all have very similar soil, sun and watering preferences. For example, parsley needs a steady supply of moisture whilst rosemary prefers dryer conditions; therefore the two herbs should not be planted in the same pot together.

Planting herbs in the same container as different types of plants- Herbs can look fantastic when planted with different types of plants such as perennials provided they have similar sun, soil and watering requirements.

Harvesting your herbs- Generally, the more you pick or prune your herbs, the more they will grow. In fact it is a good idea to regularly prune your herbs to make them more well-formed and bushier.

An example of companion planted herbs.

Chapter 3

Useful Tools

One of the advantages of container gardening is you don't need large gardening tools such as forks, rakes and spades to tend to your garden which is useful if you don't have the space to store these items.

There are some basic tools that you will need in order to make your gardening experience easier and more productive. The tools that you select will depend on many things, such as the size of your garden, what types of herbs you will be planting and what types of containers you will be growing your crops in.

The tools listed here are the some of the most basic tools you will need. You may wish to add to these tool as your need change.

Spray Packs

Spray packs are great for hydrating the leaves of more delicate herbs on a hot sunny day

Spray packs are also useful for spraying plants to prevent parasites.

Small Gardening Set

If you are using soil or potting mix as your planting medium, small gardening sets can be afford ably purchased which will make such tasks as transferring potting and container tilling mix much easier.

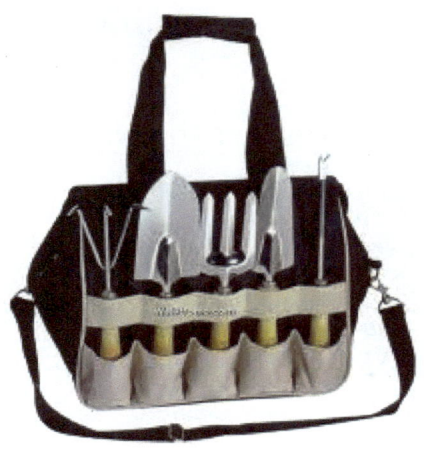

Pruning Scissors

These are great for pruning viny plants such as tomatoes, and great for harvesting herbs such as basil or rosemary.

Watering Can or Bucket

Even the most low maintenance herbs will need water sooner or later so have a watering can or some other way to easily water your plants is a good idea.

Chapter 4

The Potting Soil

Potting/Soil Mix

Where possible, your potting soil should be mixed according to the particular type of plants you are growing. Seedlings should be grown in a light, moisture-retentive, soilless mixture.

A good quality container potting mix is usually composed of peat moss, perlite and vermiculite. These soilless mixes absorb moisture very well and resist compaction, but they tend to dry out very quickly. Since they do not contain any nutrients, you may need to provide your plants with a consistent supply of fertilizer. The main advantage of a soilless mix is that it is sterile, so there is no chance of introducing pest or disease infestations.

Many container gardeners add organic components to their in container growing mix. These might include leaf mould, finished compost, composted peat, or rich garden soil. A growing medium that contains ten to twenty percent organic matter will usually not dry out as easily as a soilless mix, and will also allow you to introduce beneficial microorganisms and nutrients.

The most critical consideration when you're purchasing or blending your own potting soil is to ensure that the mix is light enough to provide adequate pore space for air, water and healthy root growth. Many months of overhead watering, without the benefit of earthworms and weather to aerate the soil, can result in compacted and unhealthy soil around the root zone. To ensure that your plants' roots have the oxygen they need for healthy growth, your potting soil should contain plenty of perlite, vermiculite, or sharp sand.

This will allow water to drain freely, and ensure that the soil is at least ten to twenty percent air.

Soil found in your outside garden is not well suited for this purpose, since it's often too heavy and may contain weed seeds and insect pests. If you are purchasing a premixed potting mix, look for a mix that is specific to indoor plants. A good medium should remain loose and drain well, yet contain enough organic matter to hold nutrients and moisture.

Most commercial organic mixes will work well, or you can create your own.
Here is a quick basic recipe for making your own potting mix:

• 1 part garden loam or topsoil
• 1 part peat moss or mature organic compost
• 1 part clean builder's sand or perlite

The organic materials in the mixture will provide structure and the sand will improve drainage. A balanced, slow-release organic fertilizer can also be added to the mix.

Chapter 5

Selecting The Right Container

Luckily, herbs aren't too fussed about what kind of pot of container you grow them in.

The only basic requirement is that the container is large enough to hold the plant and that it provided adequate drainage so the excess water can escape. If you are a beginner try to use as big a container as practicable. The reason being is that bigger pot hold more soil/medium and therefore can retain moisture for longer so you do not have to water the plant as often. If you're using potting mix/soil try to use containers which are at least ten inches wide and twelve inches deep. If you are planting larger herbs such as rosemary and intend to let them grow big, make sure you use a larger pot as this will provide the plant with extra stability. If your container does not already have drainage holes, you will need to add them. This can be done by drilling 5mm holes in the bottom or along the sides of the bottom of the container. Now line the bottom of the pot with either fly mesh or landscaping cloth; this will let the excess water escape but will allow the soil to remain in the pot.

Generally, plants in clay pots will need to be watered more often than plants in other types of containers because unglazed clay pots are porous.

Also consider the colour of the container. Darker colours tend to absorb more heat which could be a problem for some plants depending on you indoor setup. Avoid using containers made of treated wood as these may leach out chemical compounds that may be absorbed by your vegetables.

Tip

Many smaller herbs can be planted directly into larger pots and herbs which have similar sun, soil and moisture requirements can all be planted in larger pots. Make sure the potting mix is completely moist before planting by thoroughly watering the container and letting the access water run off for a few hours before planting and if you're planting your herbs from seed, always follow the instructions on the back of the seed packets.

Inspect and water the potting mix to insure the potting mix doesn't dry out.

Chapter 6

Watering Container Herbs

Because container herbs have very limited soil volume, it is very important to understand the watering requirements of any particular herbs you're growing. How much and how often you water your herbs will depend on the type of container you use, the time of year, the location of your container, how long the container has been planted and the types of herbs you are growing. This is why you should monitor your plants on a regular basis and understand the requirements of the herbs you have planted. There are many soil moisture metering devices on the market, however there is no substitute for sticking your finger about an inch into the soil to observe the moisture content of your soil.

Some herbs, such as rosemary and oregano, tolerate a certain amount of neglect and, once established, prefer dryer soils. Soil which is to moist will often harm the plant while other herbs such as parsley will require more moist living conditions and will start to wilt if left in the sun all day without adequate watering.

The best way to tell if your plant requires watering is to feel the soil. Simply stick your finger into the soil about an inch deep and if the soil feels dry water. Add enough water until the water starts to drip out of the drain holes at the base of the pot. This will indicate to you that 1) this soil has adequate moisture and 2) the soil provides sufficient drainage to allow some leaching.

By allowing the water to leach out of the base of the plant, you are flushing away any harmful build-up of soluble salts with the water. Soluble salts and other toxins can come from the water or the fertiliser and flushing it away will prevent it from building up and harming your plant. Symptoms of salt build up include burned leaf edges and damaged roots. Use the most natural source of water you can find, such as tank or well water, and avoid water which has been softened. Depending on your particular circumstances, you can water your herbs using a watering can or using a drip irrigation system.

It is worth noting that when you use soilless potting mix, if they are allowed to completely dry out, they have a tendency to shrink away from the edges of the pot which makes rewetting the mix difficult because the water tend to run down the inner edges of the pot between the sides of the pot and the mound of potting mix rather than through the potting mix, therefore leaving the centre of the potting mix dry. The problem can be remedied in two ways:

1. You can submerge the plant pot in a bucket of water until bubbles stop floating to surface of the bucket of water or,
2. Water the plant repeatedly to rehydrate the soil.

When you see moisture on the surface of the potting mix you will know that the mix is now sufficiently hydrated. To help the soil retain its moisture, thereby slightly lengthening your watering intervals, prior to planting your plant in the pot, you can also mix in one of the many hydrating gels now available. Usually these polymer-based gels look similar to granular rock salt when they dreary before they are mixed into the potting mix. When water is added to these gels, they take on a consistency similar to gelatine and will act like a reservoir of water that the plant can consume when the potting soil becomes dry. Theses gels are not designed to eliminate adequate and timely watering, however they will provide you with a slight buffer.

Chapter 7

How To Grow Herbs Indoors

There are many types of herbs that can be grown indoors without the aid of extra lighting. Here are some of the best herbs you can grow on windowsills and near natural light sources, such as glass doors and windows, and some handy hints you can implement to keep them healthy.

Cutting And Rooting

How To Plant Using a Cutting

Many herbs including oregano, rosemary, thyme and sage— are ideally propagated for indoor growing by taking a cutting from an existing outdoor plant. To do this, cut off a 4-inch section, measured back from the tip. Strip off the lower leaves and stick the stem into moist potting mix. To ensure good humidity, cover with glass or clear plastic, and keep the potting mix moist.

Light, Water And Temperature

Only plant herbs in a container that can provide adequate drainage. While most herbs like to be well watered, it can be quite easy to over-water them.
Only water when the surface of the potting mix feels dry and add sand or vermiculite to the potting mixture to encourage proper drainage.

Learn to fine-tune water, light, and temperature requirements of your plant. A plant in a terra-cotta pot in a south-facing window will need more water than one in a plastic pot in an east, or west, facing window. If the light is low, keep the temperature low.

Hold back on the water and fertilizer through winter, but when the days start getting longer, feed them with liquid seaweed or compost. Even potted soil gets compacted as you water it, so cultivate it with a little fork, then top off the surface of the soil with compost and mulch.

Spring is usually a great month for indoor plants because of all the bright light. During summer, make sure the plants are well-watered as they can quickly become dehydrated in the heat under a window.

Chapter 8

Ten Herbs You Can Grow In containers

How To Plant Herbs Using a Cutting

Many herbs, including oregano, rosemary, thyme and sage, can propagated for container growing by taking a cutting from an existing outdoor plant. To do this, cut off a 4-inch section, measured back from the tip. Strip off the lower leaves and stick the stem into moist potting mix. To ensure good humidity, cover with glass or clear plastic, and keep the potting mix moist.

Basil

Basil is a tropical plant, so if you live in a cooler climate or intend to seed it during winter, it is best to grow basil indoors under a sunny window.

Plant basil from seed in a well-drained soil enriched with aged manure or compost. Plant large-leaved basil varieties further apart as basil grows best when it has lots of air circulation. Basil needs to remain well-watered. Pinch or trim plants regularly to encourage bushy growth and trim the flowers to encourage the plant to divert its energy into producing more foliage.

Fungal diseases, such as gray mould and black spots, can attack basil, so to avoid these problems, it is best to grow basil in warmer climates and making sure not to plant basil to close to other plants.

If grown outside, pests can be controlled by picking them off of your plants. Harvest basil by picking off the largest leaves first.

Chives

Chives will grow in partial shade but prefer full sun. If you're planting chives outdoors, plant in early spring in well-draining soil containing plenty of compost or an aged manure, spacing them about ten inches apart. Chives require little care apart from watering until they are well established.

If you frequently harvest your chives it is a good idea to fertilise the plants with a liquid fertiliser such as a seaweed solution. Trim the flowers to encourage the plant to divert its energy into producing more foliage. If the pot is large enough, after three or four years, the plant will grow into a clump of smaller plants which can be divided in late winter or early spring.

Cilantro

Cilantro prefers to grow in full sun but it will tolerate light shade in the south and southwest where the sun is intense. It is best to plant cilantro in its own pot but if the container is big enough, the plants can be spaced ten  inches apart. Fall is the best time to plant cilantro, because the plants will last until the weather warms up in late spring. Plant cilantro in three-week intervals, because when the plants start to bloom, the plant will produce very little foliage.

Dill

Dill grows to a height of about two to three feet so it is important that you choose sturdy containers capable of supporting the plant. For dill, choose pots that are tall and wide.

Sow dill seeds within the first month of spring, once dill is established it will require little watering.

Fennel

Fennel prefers a fertile and well-drained soil. Position fennel plants in a sunny area. Don't position fennel near coriander or dill because they will cross-pollinate which will greatly reduce seed production.
While fennel is a perennial herb it can be grown annually in cooler climates. Because fennel easily self-sows once you plant fennel once you will see fennel popping up around the plant, which for container gardening, will
need to be uprooted. Fennel can grow up to five feet tall so you will need to sow fennel seeds in a relatively wide container which is at least ten inches deep.

Marjoram

In cooler climates, marjoram can be planted in spring when snow or frost is no longer a threat. Because marjoram takes a long time to grow, an inexperienced gardener should start with young plants rather than seeds. Marjoram prefers a fertile well-drained soil so add some compost or old manure before planting and continue to add a liquid fertiliser such as seaweed solution throughout the growing season. Trim flowers when they start to bud to encourage bushier, fuller growth.

Mint

Mint can adapt to many types of soil, however, it thrives in well fertilised, moist, well-drained soil. Mint should be positioned in a sunny place. Frequently trim mint to keep it looking good. Like basil, you will be

harvesting the leaves, so trim off the flowers to encourage the plant to divert its energy into producing more foliage. Also, trimming back the stems will encourage shorter bushier growth. Provide mint with free air circulation to prevent disease and pest infestations.

Oregano

Oregano can be planted using seed or a fresh cutting.
Use fertilizer sparingly as too much will adversely affect the taste of oregano. Provide oregano with at least six hours of direct sunlight every day. Oregano tends to lean towards the sun so make

sure that you rotate the pot to ensure even growth. Oregano

prefers drier soil, so make sure that you do not over-water the plant. Oregano planted from seed will usually take about three weeks to sprout because oregano is a Mediterranean herb, if you live in a colder climate, it is best to bring the plant indoors and place it under a sunny window as freezing temperatures will quickly kill oregano.

Parsley

The best way to plant parsley is to plant seed it indoors about two months before the end of winter. Sow parsley seed directly into pots which are at least six inches deep as Parsley has a long taproot so it will outgrow the smaller pots in no time. Keep the soil moist at all times and make sure the parsley gets plenty of sunlight.
Parsley plants will die back in their first winter and regrow again in the spring. While you may get some  parsley leaves in the second year, the plant tends to go to flower and seed quite quickly which will ruin any herb harvesting alternatively, you can harvest the seeds and dig your plants up in the late fall and just start with new seed the next spring, or allow your plants to finish their cycle.

Rosemary

If you are a beginner, it is recommended that you plant rosemary either from cutting or buy a young plant. Unless you intend to keep trimming your rosemary to keep it small, it is recommended that you plant it in a larger container. Rosemary is slow to establish itself but by the second year, its growth will really pick up speed. Rosemary will tolerate partial shade but will thrive in full sunlight. Rosemary  prefers well-drained soil with a little compost or manure mixed in. Keep the potting mix moist and allow it to dry out between waterings. Periodically fertilise your plant using a liquid fertiliser and in spring trim any dead wood from the plants.

Sage

Sage can be planted either by seed or a cutting and does well in either partial or full sun. Like rosemary, keep the potting mix moist and allow it to dry out between waterings. After about three years, sage plant will become woody and not taste as good as younger plants so it is best to pull older plants out and plant new ones.

Chapter 9

How To Make Your Own Miniature Green House At Very Little Cost

These miniature greenhouses are great for starting new plants and keeping smaller herbs in during winter and can be quickly and cheaply made using almost any disused, clear, flat sided beverage bottles, and are great in cooler climates for not only herbs but also starting almost any other plant. I personally make these using 2.5 litre drinking water bottles.

Step 1

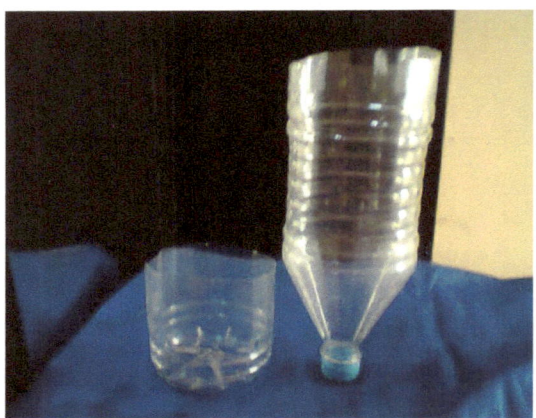

Cut bottle about four inches from the base as shown in above illustration.

Step 2

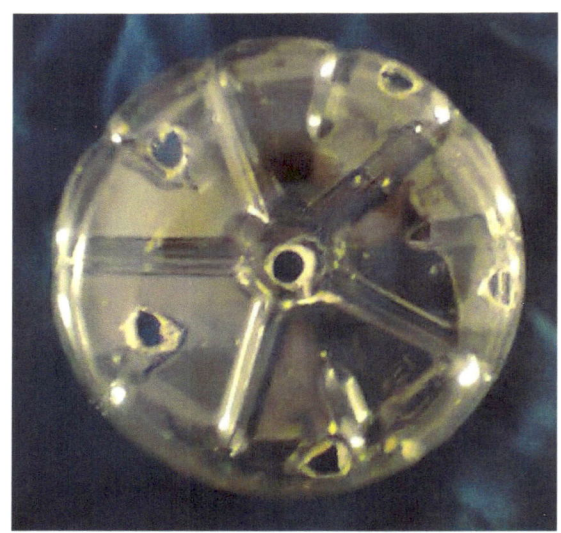

Drill about six to eight 10 millimetre holes in the base as shown in the illustration above

Step 3

After you have planted your plant in the lower half simply slide the upper (greenhouse) part into the lower (plant pot) part as shown in the illustration above.

Other Books By Dr John Stone

Backyard Pet Chickens A Beginners Guide

How To Raise Hens In A Small Suburban Yard

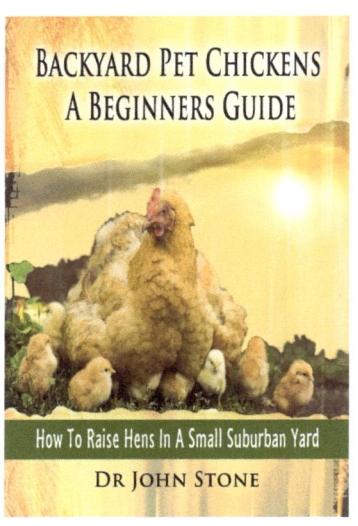

You are what you eat. But, Dr. John Stone, extends that axiom further--you are what you eat, eats. Backyard Pet Chickens A Beginners Guide is a shocking report of what's happened to our food and a comforting solution in how to help control the quality of what we eat by getting into organic chicken production. In an easy, step-by-step process, the reader is given a turn-key experience to starting a backyard chicken coop. Nothing is left out, not even observance of state and local laws—Dr. Stone covers it all—which is what makes this book stand out from all the others: it gives the reader honest, accurate information that can transform their lives while improving the quality of their diet and health.

Indoor Gardening Made Easy

How To Grow Herbs And Vegetables In Your House

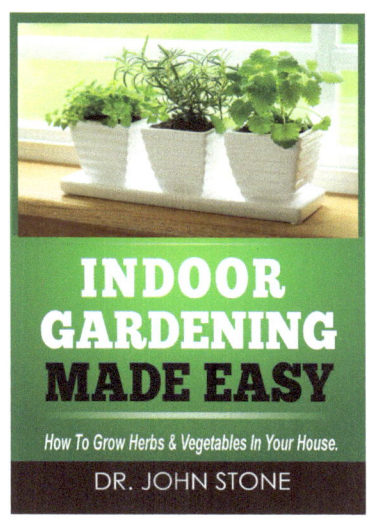

Gardening is an activity that is good for both you and your environment; it is relaxing and leaves you with a sense of achievement. With people increasingly living in apartments and in urban areas, they often assume that a garden is not even an option for them. There's no need to miss out on the joy and health benefits gardening brings just because you may not have an outside space. Whether you live in an urban loft, a studio apartment, or a traditional house, "Indoor Gardening Made Easy" will be your guide to bringing a little bit of outside, inside.

Organic Pest Control Made Easy

How To Naturally Keep Your Home, Garden & Food Pest Free

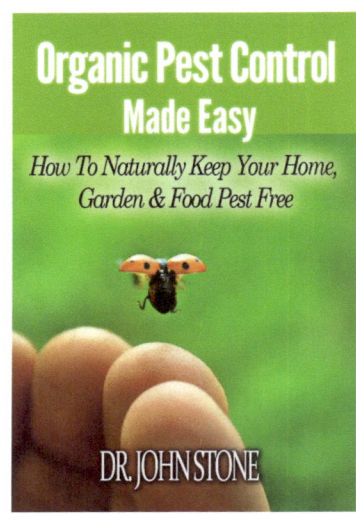

The truth is, apart from the possible long term health problems, modern factory farming and chemical pesticides are actually less effective in the long term than organic gardening.

With pest control in the home and garden, nature provides all of the solutions if you know where to look.

This book will teach you everything you need to know to not only protect your home and garden from pests but to also protect your home, garden and food from pesticides.

Organic Composting Made Easy

How To Create Natural Fertilizer At Home

In this day and age it's difficult to know what is really in your food, even vegetables labelled as "organic" can contain artificial substances and are often grown in fertilizers that can harm both the earth and the plants themselves.
Compost has been made from organic matter that has decomposed; it is the best, most environmentally friendly fertilizer for your plants.
Whether you want to have an organic farm so that you know exactly where your fruits and vegetables come from or you want a thriving garden, composting is the way to go and this book will help you along the way.

www.ingramcontent.com/pod-product-compliance
Lightning Source LLC
Chambersburg PA
CBHW041151180526
45159CB00002BB/785